MW01001924

THE INVENTOR SAYS

Also available in the
Words of Wisdom series:

The Architect Says
Laura S. Dushkes

The Designer Says
Sara Bader

The Filmmaker Says
Jamie Thompson Stern

The Chef Says
Nach Waxman and Matt Sartwell

The Musician Says
Benedetta LoBalbo

the INVENTOR *says*

Quotes, Quips, and Words of Wisdom

compiled & edited by Kevin Lippert

Princeton Architectural Press, New York

We're all, in one way or another, inventors. While we may not be cobbling together the next world-changing communication device in the basement with wire, soldering iron, duct tape, and breadboard, we face problems every day that require innovation, know-how, and *aha!* moments, whether devising a system for attaching a tail to a Halloween costume; figuring out a better way to organize and store tools, kitchen utensils, or cables; fixing a lawn mower with materials at hand; or just trying to simplify and make easier the challenges and chaos of everyday life. Possibly because we know firsthand how demanding even these "quick and simple" inventions can be – and rewarding when they succeed – we hold inventors in special esteem, with an extra appreciation of the conceptual leaps, genius, or sometimes plain dumb luck these pioneers, including many legends, needed to improve their lives and ours, and, in many cases, to change the world as we know it.

Compiling quotes for this latest volume in our Words of Wisdom series, I was struck by how many are from white men of the nineteenth and twentieth centuries, mostly American. Before we jump to conclusions about

Yankee ingenuity or a male penchant for tinkering, this is most likely a reflection of the role of women in our recent past, the lack of educational opportunities in science and math, my own cultural bias, and the emergence of journalism and publishing in those years, which afforded the opportunity for (mostly male) inventors to offer inspirational, often catchy advice bordering on self-promotion – something people like Thomas Edison and Henry Ford took seriously. Ts'ai Lun undoubtedly had something to say after inventing paper in China around AD 105, but no scribe was there to write it down for us. Edison and Ford, on the other hand, rarely missed a chance to quip about what they did, and how we might do things, too.

The other striking thing is how many inventors' quotes deal with failure, reinforcing the old sports cliché that it's not how many times you get knocked down, but how many times you get up that counts. If there's a single message to glean from this collection of 150 quotes, it's that success is almost always built on the foundation of false starts, something we all should remember the next time the costume tail falls off or our cable organizers tear out of the wall.

I JUST WANT THE FUTURE TO HAPPEN FASTER.

Nolan Bushnell (1943–)

Inventor of Pong

But still try, for who knows what is possible?

Michael Faraday (1791–1867)

Inventor of the electric induction motor

If you are curious, you'll find the puzzles around you. If you are determined, you will solve them.

Ernő Rubik (1944–)

Inventor of the Rubik's Cube

The greatest thing in our favor was growing up in a family where there was always much encouragement to intellectual curiosity.

Orville Wright (1871–1948)

Coinventor of the first successful airplane

MY INVARIABLE QUESTION ON RECEIVING ANY NEW TOY WAS: "MAMMA, WHAT IS INSIDE OF IT?"

Charles Babbage (1791–1871)

Inventor of the automatic mechanical calculator

I was a very creative child.
I watched my mother sewing and making patterns.
I imitated what she did.
I had all these paper dolls, and I remember just lying on the floor and drawing, making costumes for these paper dolls.

Stephanie Kwolek (1923–2014)

Inventor of Kevlar

I was always making things for my brothers; did they want any thing in the line of playthings, they always said, "Mattie will make them for us." I was famous for my kites; and my sleds were the envy and admiration of all the boys in town.

Margaret Knight (1838–1914)

Inventor of a machine to produce paper grocery bags

ONE OF MY GREATEST DESIRES WHEN I WAS ONLY TEN YEARS OLD WAS TO HAVE A TYPEWRITER.
ONE OF MY AUNTS GAVE ME A TOY TYPEWRITER WITH WHICH YOU COULD ACTUALLY TYPE. IT WAS A RUBBER STAMP AFFAIR.... SOMEHOW THE GRAPHIC ARTS INTERESTED ME ALL THROUGH THAT PERIOD.

Chester Carlson (1906–68)

Inventor of the xerographic process

I wished for nothing more than to possess a small printing press, *and thus to be the composer, printer, and publisher of my own productions. Had I then possessed sufficient means to gratify this wish, I should never, perhaps, have been the inventor of the Lithographic art.*

Alois Senefelder (1771–1834)

Inventor of lithography

I was very fond of machinery, and of watching it when in motion; and if ever I was absent from meals, I could probably have been found at the flour mill at the other end of the village, where I passed many hours.

Sir Henry Bessemer (1813–98)

Inventor of the first method for mass-producing steel

When I was a little kid, I...made a sketch of what I thought an atom bomb would look like.... I had that sketch for a number of years, and when pictures of the atom bomb were published many years later, I found that my sketch wasn't far off. The atom bomb was pretty easy to invent, once you understood the basic principles.

Mary Spaeth (1938–)

Inventor of the tunable dye laser

When I was a teenager, my father was a physician; he took me to watch surgery, and it occurred to me at the time . . . that it was very cumbersome, some of the things they were doing by hand, so ***I invented some surgical stapling instruments, and started building them as a child.***

Robert Jarvik (1946–)

Inventor of the first permanent artificial human heart

There are maybe three inventions I have that I rank as my top inventions that I'm most proud of. The robot I built in high school, the memory-protected circuitry for the Galileo [spacecraft], and the Super Soaker.

Lonnie Johnson (1949–)

Inventor of the Super Soaker

I was...the smallest kid in the class all through high school—not especially coordinated and certainly not the football-player type. All this made me socially uncomfortable and probably helped guide me toward model airplanes as a hobby. **Looking back, I'm glad I had those limits.**

Paul MacCready (1925–2007)

Inventor of the Gossamer Condor human-powered aircraft

I am not the kind of person to enjoy going out golfing or fishing, or even partying.

Chester Carlson (1906–68)

Inventor of the xerographic process

An outdoor life is necessary to health and success, especially in a youth.

Alexander Graham Bell (1847–1922)

Inventor of the telephone

ATHLETICS AND GYMNASTICS...
WERE LOOKED ON AS
A DEVICE OF THE DEVIL
TO LEAD YOUNG MEN ASTRAY.
I FELT, HOWEVER, THAT IF
THE DEVIL WAS MAKING
USE OF THEM TO LEAD YOUNG
MEN, IT MUST HAVE SOME
NATURAL ATTRACTION, AND
THAT IT MIGHT BE USED
TO LEAD TO A GOOD END AS
WELL AS TO A BAD ONE.

James Naismith (1861–1939)

Inventor of basketball

The difference in character and future usefulness between the self-educated and the college-pampered person is the difference between the baby who can sit up and feed himself and **the one who has to be held and have the food spooned into him.**

Earl Silas Tupper (1907–83)

Inventor of Tupperware

If I had been technically trained, I would have quit or probably never would have begun.

King Camp Gillette (1855–1932)

Inventor of the first safety razor with a disposable blade

ANY YOUTH WHO MAKES SECURITY HIS GOAL SHACKLES HIMSELF AT THE VERY START OF LIFE'S RACE.

Clarence Birdseye (1886–1956)

Inventor of a quick-freezing process for food

Confidence is sometimes rooted in the unpleasant, harsh aspects of life, and not in warmth and safety.

An Wang (1920–90)

Inventor of magnetic-core memory

You read everything —that's part of the job. You accumulate all this trivia, and ***hope that someday maybe a millionth of it will be useful.***

Jack Kilby (1923–2005)

Coinventor of the integrated circuit

THE WORLD ISN'T BLACK AND WHITE. IT'S GRAY SCALE. AS AN INVENTOR, YOU HAVE TO SEE THINGS IN GRAY SCALE.

Steve Wozniak (1950–)

Cofounder of Apple Inc.

I was then eighteen.
While working on an experiment,
I failed, and was about to throw
a certain black residue away when
I thought it might be interesting.
The solution of it resulted
in a strangely beautiful color.
You know the rest.

William Henry Perkin (1838–1907)

Inventor of mauveine, the first synthetic organic chemical dye

At 3M we're a bunch of ideas. **We never throw an idea away** because you never know when someone else may need it.

Arthur Fry (1931–)

Coinventor of the Post-it Note

I learned to observe people in their daily lives, and, every once in a while, **I identify a need.**

Ruth Handler (1916–2002)

Creator of the Barbie doll

The primary idea in all of my work was to help the farmer and fill the poor man's empty dinner pail.... **My idea was to help the "man farthest down";** this is why I have made every process just as simple as I could to put it within his reach.

George Washington Carver (ca. 1864–1943)

Developer of sweet potato and peanut products

A SERIOUS MISMATCH
HAS DEVELOPED
BETWEEN TECHNOLOGY
AND OUR
SOCIAL INSTITUTIONS...
**INVENTIVE MINDS OUGHT
TO CONSIDER
SOCIAL INVENTIONS
AS THEIR
FIRST PRIORITY.**

Dennis Gabor (1900–79)

Inventor of holography

If some day they say of me that I have contributed something to the welfare and happiness of my fellow man, I shall be satisfied.

George Westinghouse (1846–1914)

Inventor of the railway air brake

*OUR PLAN IS TO
LEAD THE PUBLIC WITH
NEW PRODUCTS
RATHER THAN ASK THEM
WHAT KIND OF PRODUCTS
THEY WANT.*
***THE PUBLIC DOES NOT
KNOW WHAT IS POSSIBLE,
BUT WE DO.***

Akio Morita (1921–99)

Cofounder of Sony and developer of the Walkman

I think about things we need and want **before we even know we need and want them.**

Joy Mangano (1956–)

Inventor of the Miracle Mop

A person without a practical end in view becomes a crank or an idiot. . . . Perseverance is first, but practicability is chief.

Alexander Graham Bell (1847–1922)

Inventor of the metal detector

Resolve to perform what you ought; perform without fail what you resolve.

Benjamin Franklin (1705–90)

Inventor of the lightning rod

Perhaps some will say
that in the discovery and
solution of a problem
it is of no little assistance
first to be conscious
in some way that the
goal is a real one, and to
be sure that one is not
attempting the impossible.

Galileo Galilei (1564–1642)

Inventor of the telescope

DON'T UNDERTAKE A PROJECT UNLESS IT IS MANIFESTLY IMPORTANT AND NEARLY IMPOSSIBLE.

Edwin Land (1909–91)

Inventor of the Polaroid camera

Sooner or later...
our attitude should be not that
people should have to learn
how to code for the computer
but rather **the computer should
learn how to respond to people,**
because I figured we weren't
going to teach the whole
population of the United States
how to write computer code.

Grace Murray Hopper (1906–92)

*Codeveloper of the COBOL
business programming language*

JUST AS AN AUTOMOBILE DEALER DOES NOT ASK A CUSTOMER TO STUDY AUTOMOTIVE ENGINEERING, A COMPUTER FIRM SHOULD NOT DEMAND THAT ITS CUSTOMERS LEARN ABOUT COMPUTERS.

An Wang (1920–90)

Inventor of magnetic-core memory

The job of computers and networks is to **get out of the way,** to not be seen.

Sir Tim Berners-Lee (1955–)

Inventor of the World Wide Web

Whatever you do, the most important [thing] is ease of use; second thing is ease of use; third is ease of use.

Karlheinz Brandenburg (1954–)

Coinventor of the MP3

The science of Nature has been already too long made only a work of the brain and the fancy: **It is now high time that it should return to the plainness and soundness of observations** on material and obvious things.

Robert Hooke (1635–1703)

Inventor of the balance spring

Those who study the ancients and not the works of Nature are stepsons and not sons of Nature, the mother of all good authors.

Leonardo da Vinci (1452–1519)

Inventor of the helicopter

I get a lot from nature. I get inspired by the low leaf on the lean tree, grain of rice, looking at animals on the Serengeti, the cheetah, the giraffe and so on. ***We live on a truly magical planet*** *and that's the source of my inspiration.*

Jelani Aliyu (1966–)

Designer of the Chevrolet Volt hybrid electric car

Among the ignorant,
there have been many who,
having something to put
into words, have in the end
been unable to express
their feelings. I have been
distressed by this, and
have newly designed a script
of twenty-eight letters,
which I wish to have everyone
practice at their ease....
Even the sound of the winds,
the cry of the crane
and the barking of the dog
—all may be written.

Sejong the Great (1397–1450)

Korean emperor and inventor
of the Hangul alphabet

THE FORCES OF NATURE CANNOT BE ELIMINATED BUT THEY MAY BE **BALANCED ONE AGAINST THE OTHER.**

Ferdinand von Zeppelin (1838–1917)

Inventor of the zeppelin airship

The history of civilization is in the main the story of man's progress towards independence of the weather.

Sir Robert Watson-Watt (1892–1973)

Developer of radar technology

What led me to build the first snowboard [was] not being able to skateboard on an icy street. So after that, it was just ***twelve months a year of boarding,*** *whether it be skateboarding, snowboarding, or surfing.*

Tom Sims (1950–2012)

Developer of the snowboard

I was fed up with hooks and eyes, rusting metal and everything pertaining to the fastener.

Gideon Sundback (1880–1954)

Inventor of the zipper

A bird is an instrument working according to mathematical law, an instrument which is within the capacity of man to reproduce with all its movements.... Such an instrument constructed by man is ***lacking in nothing except the life of the bird,*** *and this life must be supplied from that of man.*

Leonardo da Vinci (1452–1519)

Inventor of the parachute

THE BIRDS' WINGS ARE UNDOUBTEDLY
VERY WELL DESIGNED INDEED,
BUT IT IS NOT ANY EXTRAORDINARY
EFFICIENCY THAT STRIKES
WITH ASTONISHMENT BUT RATHER
THE MARVELOUS SKILL
WITH WHICH THEY ARE USED....
THE SOARING PROBLEM
IS APPARENTLY NOT SO MUCH ONE
OF BETTER WINGS
AS OF BETTER OPERATORS.

Wilbur Wright (1867–1912)

Coinventor of the first
successful airplane

I experimented in my mother's flat, using wood, rubber bands and paper clips to make a prototype.... I knew it was revolutionary. The moment I started twisting the sides, I could see it was a proper puzzle—but what I didn't know was whether it could be solved. It took me weeks: there are 43 quintillion permutations!

Ernő Rubik (1944–)

Inventor of the Rubik's Cube

I'VE REALIZED MY DREAM THAT NOODLES CAN GO INTO SPACE.

Momofuku Ando (1910–2007)

Inventor of instant noodles

With lettering, an artist never corrects by erasing but always paints over the error, so **I decided to use what artists use.** I put some tempera water-based paint in a bottle and took my watercolor brush to the office. And I used that to correct my typing mistakes.

Bette Nesmith Graham (1924–1980)

Inventor of Liquid Paper

I have
no particular
occupation,
but
I know
a lot about
feathers.

Susan Hibbard (dates unknown)

Inventor of an improved feather duster

[The Trapper Keeper] was no accident. **It was the most scientific and pragmatically planned product ever** in that industry.

E. Bryant Crutchfield (ca. 1937–)

Inventor of the Trapper Keeper school binder

I'm a big planner, but I like to have my plans written in water.

Nolan Bushnell (1943–)

Inventor of Pong

I DON'T THINK NECESSITY IS THE MOTHER OF INVENTION. INVENTION... ARISES DIRECTLY FROM IDLENESS, POSSIBLY ALSO FROM LAZINESS. **TO SAVE ONESELF TROUBLE.**

Agatha Christie (1890–1976)

Author

NAME THE GREATEST OF ALL THE INVENTORS: ACCIDENT.

Mark Twain (1835–1910)

Author

If [Edison] had a needle to find in a haystack
he would not stop to reason where it was
most likely to be, but would proceed at once,
with the feverish diligence of a bee,
to examine straw after straw until he found
the object of his search.... **His method
was inefficient in the extreme,** for an immense
ground had to be covered to get anything
at all unless blind chance intervened.

Nikola Tesla (1856–1943)

*Inventor of the alternating-current
(AC) electricity system*

I NEVER DID ANYTHING WORTH DOING BY ACCIDENT.

Thomas Edison (1847-1931)

Inventor of the monofilament electric light bulb

Isn't everybody a procrastinator? The stimulus often comes from the deadline.

Paul MacCready (1925–2007)

Inventor of the Gossamer Condor human-powered aircraft

Give them the third best to go on with; the second best comes too late, the best never comes.

Sir Robert Watson-Watt (1892–1973)

Developer of radar technology

I do not rush
into constructive work.
When I get an idea,
I start right away
to build it up in my mind.
I change the structure,
I make improvements,
I experiment,
I run the device
in my mind.

Nikola Tesla (1863–1943)

Inventor of the steam-powered mechanical oscillator

Every design problem
I've ever solved
started with my ability
to visualize and see
the world in pictures...
before I attempt
any construction,
I test run the equipment
in my imagination.

Temple Grandin (1947–)

Inventor of livestock
handling equipment

I am an inspirational inventor.
I get a complete picture
in my mind of what the invention
will be like when it is finished
and then set to work to get my
model-maker to create a model
to fit my mind's picture.
Inventing is really easy;
it's the development work
that is heartbreaking.

Beulah Louise Henry (1887–1973)

Inventor of the bobbin-free
sewing machine

I almost never have an idea of what my inventions are going to look like in their final form.

Wilson Greatbatch (1919–2011)

Inventor of the implantable pacemaker

The component parts of all new machines may be said to be old.... Therefore, the mechanic should sit down among levers, screws, wedges, wheels, etc. like a poet among the letters of the alphabet, considering them as the exhibition of his thoughts; in which a new arrangement transmits a new idea to the world.

Robert Fulton (1765–1815)

Developer of the first commercially viable steamboat

WHAT I DID WAS IN SOME WAYS INEVITABLE. ALL MY LIFE, ALL THE LITTLE STEPS, ALL THE RIGHT EXPERIENCES IN TV, VIDEO, CIRCUITS, BUILDING MY OWN TERMINALS, WORKING AT ATARI, WORKING AT HEWLETT-PACKARD — ALL THESE THINGS CONVERGED. IF YOU COMBINED THEM ALL AT THAT POINT IN TIME, THEY WERE DEFINITELY GOING TO BE THE APPLE II.

Steve Wozniak (1950–)

Cofounder of Apple Inc.

Progress happens when all the factors that make for it are ready, and then it is inevitable. To teach that a comparatively few men are responsible for the greatest forward steps of mankind is the worst sort of nonsense.

Henry Ford (1863–1947)

Inventor of the moving assembly line

Ideas don't come out of a collective. People who say an invention is in the air or is a product of the times simply don't understand the process.

Stanford Ovshinsky (1922–2012)

Inventor of the nickel-metal hydride battery

VERY OFTEN I'VE FOUND THAT FRESH, NOVEL IDEAS IN A PARTICULAR DISCIPLINE DO NOT ALWAYS COME FROM THE EXPERTS IN THAT FIELD, BUT FROM THOSE IN OTHER FIELDS.... INVENTION IS INCREASINGLY INTERDISCIPLINARY WORK. IT INVOLVES TEAMS OF PEOPLE. **THERE ARE RELATIVELY FEW SIGNIFICANT INVENTIONS THAT CAN BE CREATED BY ONE OR TWO PEOPLE.**

Raymond Kurzweil (1948-)

Principal inventor of the flatbed scanner

With all the needed emphasis on leadership, organization, and teamwork, the individual has remained supreme—of paramount importance. It is in the mind of a single person that creative ideas and concepts are born.

Mervin Kelly (1894–1971)

Codeveloper of the transistor

We designed the house with my lab right off the living room so that I wouldn't have to say good-bye to the family when I wanted to fool around. The sliding panels on the bookshelf actually open into my lab so that it's connected to the living room.

Jacob Rabinow (1910–99)

Inventor of the optical character recognition machine

Be alone, that is the secret of invention; be alone, that is when ideas are born.

Nikola Tesla (1863–1943)

Inventor of the alternating-current (AC) motor

I have a number of patents, so that I am assessed as being an inventor by others. They invite me to join inventors' societies. I don't do so because they don't impress me very much, and I don't think an inventor is very good en masse. **Invention is something you have to do by yourself.**

R. **Buckminster Fuller** (1895–1983)

Inventor of the geodesic dome and other architectural structures

IF YOU WORK ON YOUR OWN AS I DO, NOBODY KNOWS IF YOU MAKE A MISTAKE.

Roman Szpur (1916–2008)

Inventor of the door-mounted burglar alarm

The man who is certain he is right is ***almost sure to be wrong,*** *and he has the additional misfortune of inevitably remaining so.*

Michael Faraday (1791–1867)

Inventor of the electric generator

I HAVE NO DOUBT OF SUCCESS —NONE.

Gail Borden (1801–74)

Inventor of condensed milk

I believe that I can compete with anyone in architecture, and in the construction of both public and private monuments, and in the building of canals. I am able to execute statues in marble, bronze, and clay; ***in painting I can do as well as anyone else.***

Leonardo da Vinci (1452–1519)

Inventor of the armored car

It's said that God owes a lot to Johann Sebastian Bach. I would like it said that French people owe a lot to Moreno.

Roland Moreno (1945–2012)

Inventor of the smart card

You could walk into any store in America and buy a waterbed. How groovy is that?

Charles Prior Hall (ca. 1943–)

Inventor of the waterbed

If you buy my lamp, you won't need drugs.

Edward Craven Walker (1918–2000)

Inventor of the lava lamp

WE ARE CONTINUING OUR EXPERIMENTAL AND DEVELOPMENTAL WORK CONFIDENT IN THE BELIEF THAT THE REAL POSSIBILITIES OF SLICED BREAD HAVE SCARCELY BEEN SCRATCHED.

Otto Frederick Rohwedder (1880–1960)

Inventor of the automatic bread-slicing machine

The flush toilet may have been the most civilized invention ever devised, but the remote control is the next most important. It's almost as important as sex.

Eugene Polley (1915–2012)

Inventor of the wireless remote control

You have to be able to **take your ego and throw it away.** Everybody has a better idea than you.

Emanuel Logan Jr. (1933–2005)

Inventor of the bulletproof Plexiglas shields for bank tellers

IF SOMEBODY HAS IMAGINATION AND THINKS UP NEW CONCEPTS AND NEW IDEAS, WE ARE VERY APT TO PUT A COMMITTEE AROUND THEM AND CUT THEM DOWN TO SIZE.... WE ARE AFRAID OF GIANTS, SO WE ARE NOT GROWING ANY. WE NEED A FEW GIANTS.

Grace Murray Hopper (1906–92)

Inventor of the first computer compiler

If I have seen further, it is by standing on the shoulders of giants.

Sir Isaac Newton (1643–1727)

Developer of calculus

I was able to stand on
the shoulders of those women
who came before me,
and women who came after me
were able to stand on mine.

Christine Darden (1942–)

Developer for NASA of high-lift wing designs to minimize sonic boom

We hear much of chivalry of men towards women; but let me tell you, gentle reader, it vanishes like dew before the summer sun when one of us comes into **competition with the manly sex.**

Martha J. Coston (1826-1904)

Inventor of the Coston signal flare

NO MAN RESPECTS WOMEN MORE THAN I DO, BUT NO WOMAN SINCE CREATION EVER INVENTED ANYTHING, AND NO FEMALE EVER WILL.

Gail Borden (1801–74)

Inventor of condensed milk

THERE'S JUST SOMETHING ABOUT A BIG COMPANY THAT **SNUFFS OUT THE ENTREPRENEURIAL SPIRIT** THAT'S NEEDED IN A WHOLLY NEW FIELD.

Gordon Gould (1920–2005)

Inventor of the laser

Early in life, I discovered that most of the good things I had done happened when the boss was out of town.

Robert Gundlach (1926–2010)

Developer of numerous Xerox technologies

People have sometimes asked me whether I am upset that I have not made a lot of money from the Web. In fact, I made some quite conscious decisions about which way to take my life. . . . What is maddening is **the terrible notion that a person's value depends on how important and financially successful they are,** and that that is measured in terms of money.

Sir Tim Berners-Lee (1955–)

Inventor of the World Wide Web

LET ANY MAN PLACE
HIMSELF IN MY SITUATION
AND THEN ASK HIMSELF
IF HE HAS NOT A RIGHT
TO CONVERT HIS LABORS
INTO FAME AND EMOLUMENT,
FOR WHAT OTHER OBJECTS
DO MEN LABOR?

Robert Fulton (1765–1815)

*Developer of the first
commercially viable steamboat*

Invention was one chance to start with nothing and end up with a fortune.

Chester Carlson (1906–68)

Inventor of the xerographic process

It's not always super-well paid, but, let's face it, you can only wear one suit at a time, and I hate wearing suits.

Trevor Graham Baylis (1937-)

Inventor of the wind-up radio

My object in life
is not simply to make
money for myself
or to spend it on myself
in dressing or running
around in an automobile.
**But I love to use
a part of what I make
in trying to help others.**

Madame C. J. Walker (1867–1919)

*Developer of hair products
for African Americans*

I always invent to obtain money to go on inventing.

Thomas Edison (1847–1931)

Inventor of the phonograph

I've been fired, I've been called a loser
and just all kinds of things, so
now I'm getting acclimated to that.
But **the road along there, it was
cold and lonesome.** There's a lot of times
you wonder that, God, maybe I'm
just really wrong, maybe I've got some
real aberration in my subconscious
and I'm really way off.

Douglas Engelbart (1925–2013)

Inventor of the computer mouse

IF YOU CALL SOMEONE AN INVENTOR, YOU'RE IMPLYING THAT HE'S JUST MONKEYING AROUND. AS A MATTER OF FACT, MOST INVENTIONS DO FAIL. AN INVENTOR MAY TRY HUNDREDS OF THINGS THAT DON'T WORK, AND THAT GIVES MOST PEOPLE THE IMPRESSION THAT HE IS SOMEWHAT CRAZY.

Marvin Camras (1916–95)

Inventor of magnetic tape recording

*I have labored hard against the strong current of disappointment, which has been threatening to carry us down the cataract, but **I have labored with a shattered oar.***

Eli Whitney (1765–1825)

Inventor of the cotton gin

Though from my earliest years I had been uniformly deluded by hope, I still continued to yield myself up to its allurements.

Alois Senefelder (1771–1834)

Inventor of lithography

Mankind, generally, are not to be depended on, and the best workmen I can find are incapable of directing. Indeed, there is no branch of the work that can proceed well, scarcely for a single hour, unless I am present.

Eli Whitney (1765–1825)

Inventor of the milling machine

I am extremely indolent,
cannot force workmen
to do their duty, have been
cheated by undertakers
and clerks, and am unlucky
enough to know it.
The work done is slovenly,
our workmen are bad,
and I am not sufficiently
strict... In short I find
myself out of my sphere
when I have anything to do
with mankind.

James Watt (1736–1819)

*Inventor of the Watt
steam engine*

A PHYSICIST IS ONE WHO'S CONCERNED WITH THE TRUTH, AND AN ENGINEER IS ONE WHO IS CONCERNED WITH GETTING THE JOB DONE AND NOT WITH THE TRUTH. AN ENGINEER COULD, IF NECESSARY, GO OUT WITH A GUN AND HOLD UP SOMEBODY ON THE STREET TO GET THE JOB DONE—A GOOD ENGINEER WOULD PROBABLY DO IT, YOU SEE—A PHYSICIST WOULD NOT. ***I'M MAKING A SLIGHT EXAGGERATION.***

J. Presper Eckert (1919–95)

Coinventor of the ENIAC computer

In some sort of crude sense which no vulgarity, no humor, no overstatement can quite extinguish, the physicists have known sin; and this is a knowledge which they cannot lose.

J. Robert Oppenheimer (1904–67)

Supervisor of the development of the atomic bomb

TRUTH HAS BEEN THE OBJECT OF MY SEARCH, AND I AM NOT CONSCIOUS OF EVER HAVING TURNED ASIDE IN MY ENQUIRIES FROM ANY FEAR OF THE CONCLUSIONS TO WHICH THEY MIGHT LEAD.

Charles Babbage (1791–1871)

Inventor of the locomotive pilot

THE ONLY REGRET I HAVE ABOUT THE TRANSISTOR IS ITS USE FOR ROCK AND ROLL.

Walter Brattain (1902–87)

Coinventor of the transistor

I don't take responsibility for couch potatoes. They really should exercise.

Robert Adler (1913–2007)

Inventor of the improved wireless remote control

I'm sorry. Our intentions were good.

Ethan Zuckerman (1972–)

Inventor of the pop-up ad

I got in the scanner, and we didn't get any signal. It was a profound disappointment. We hypothesized that the scan failed because, frankly, **I was just too fat for the coil.**

Raymond Damadian (1936–)

Inventor of the Magnetic Resonance Scanning Machine, precursor to the MRI

The rate at which a person can mature is directly proportional to the embarrassment he can tolerate.

Douglas Engelbart (1925–2013)

Developer of hypertext

I see a man must either resolve to put out nothing new, or to **become a slave to defend it.**

Sir Isaac Newton (1643–1727)

Inventor of the reflecting telescope

To be widely read in English-speaking countries ***think of the most stupid student you have ever had,*** *then think how you would explain the subject to him.*

Arthur Holmes (1890–1965)

Developer of the geological time scale

Long experience has taught me this about the status of mankind with regard to matters requiring thought: **the less people know and understand about them, the more positively they attempt to argue concerning them,** while on the other hand to know and understand a multitude of things renders men cautious in passing judgment upon anything new.

Galileo Galilei (1564–1642)

Inventor of the early thermometer

Thinking is the hardest work there is, which is the probable reason why so few engage in it.

Henry Ford (1863–1947)

Inventor of the moving assembly line

When you can measure what you are speaking about, and express it in numbers, ***you know something about it.***

Sir William Thomson, Lord Kelvin (1824–1907)

Inventor of the Kelvin scale of temperature

IN MATHEMATICS YOU DON'T UNDERSTAND THINGS. YOU JUST GET USED TO THEM.

John von Neumann (1903–57)

Developer of game theory

I seem to have spent much more of my life **not solving structures** than solving them.

Dorothy Crowfoot Hodgkin (1910–94)

Inventor of X-ray crystallography

An interesting life can’t be all violins and flowers. When you lose a game of chess, you don’t go and jump off a bridge, you reset the pieces and do it again.

Nolan Bushnell (1943–)

Inventor of Pong

I FIND THAT NOTHING BUT VERY CLOSE AND INTENSE APPLICATION TO SUBJECTS OF A SCIENTIFIC NATURE NOW SEEMS TO KEEP MY IMAGINATION FROM RUNNING WILD.

Ada Lovelace (1815–52)

Developer of the first computer program

Concentrate all your thoughts upon the work at hand. The sun's rays do not burn until brought to a focus.

Alexander Graham Bell (1847–1922)

Inventor of the telephone

Every great and deep difficulty bears within itself its own solution.

Niels Bohr (1885–1962)

Developer of quantum theory

FRICTION BRINGS FORTH THE SPARKS OF LATENT FIRE.

Robert Fulton (1765–1815)

Developer of the first commercially viable steamboat

WHATEVER I HAVE
ACCOMPLISHED IN LIFE
I HAVE PAID FOR IT
BY MUCH THOUGHT
AND HARD WORK.
**IF THERE IS ANY EASY WAY,
I HAVEN'T FOUND IT.**

Madame C. J. Walker (1867–1919)

*Developer of hair products
for African Americans*

I came up with the idea in my sleep. To be honest, I'm a lazy bum and my productivity is on the feeble side. I'm jealous, spendthrift, a total couch potato, and absent-minded.

Roland Moreno (1945–2012)

Inventor of the smart card

WE TRY EVERYTHING, BUT WE TRY THE RIGHT THING FIRST!

Edwin Land (1909–91)

Inventor of the Polaroid camera

I was always afraid of things that worked *the first* time.

Thomas Edison (1847–1931)

Inventor of the moving-picture camera

If you're always waiting for
that wonderful breakthrough,
it's probably never going to happen.
Instead, what you have to do is
keep working on things.

Marcian Hoff (1937–)

Coinventor of the microprocessor

There are no big steps, there are all little steps.

Gregory Goodwin Pincus (1903–67)

Coinventor of the combined oral contraceptive pill

THE NOVELTY OF INVENTING PROGRAMS WEARS OFF AND DEGENERATES INTO THE DULL LABOR OF WRITING AND CHECKING PROGRAMS. THE DUTY NOW LOOMS AS AN IMPOSITION ON THE HUMAN BRAIN.

Grace Murray Hopper (1906–92)

Inventor of the first computer compiler

I've always loved writing software. I think part of that love stems from the fact that I get to boss computers around. Unlike my kids, there's no hesitation or talking back. ***If you ask a computer to do something, it will hang on your every word.***

Lisa Seacat DeLuca (1983–)

IBM's most prolific female inventor, with more than 150 patents

I'm not interested in developing a powerful brain. All I'm after is **just a mediocre brain,** something like the President of the American Telephone and Telegraph Company.

Alan Turing (1912–54)

Inventor of the bombe World War II code-breaking machine

It's easy to get the steering to work 99% of the time. But 99% is not good enough. 1% of steering into a wall is not ...not good.

Elon Musk (1971–)

Developer of electric car and space travel technologies

I WAS TERRIBLY EXCITED AND PROUD, BUT AFTER A FEW DAYS I WAS INFORMED, VERY POLITELY, THAT MY DEMONSTRATION HAD BEEN EXTREMELY INTERESTING, BUT THAT **IT MIGHT BE BETTER IF I WERE TO SPEND MY TIME ON SOMETHING "A LITTLE MORE USEFUL."**

Vladimir K. Zworykin (1888–1982)

Inventor of early television systems

I learned very early on that you can use negative reactions as stepping stones. I found that they gave me even more inspiration to say, "Well, dammit, I'll show them."

Nathaniel Wyeth (1911–90)

Inventor of the plastic soda bottle

99% OF SUCCESS IS BUILT ON FAILURE.

Charles Kettering (1876–1958)

Inventor of the electric car starter and leaded gasoline

The key to doing something right may lie in the feedback you get from doing something wrong.

An Wang (1920–90)

Inventor of magnetic-core memory

When it is all done
and is a success, I can't
bear the sight of it.
I haven't used a telephone
in ten years, and
I would go out of my way
any day to miss
an incandescent light.

Thomas Edison (1847–1931)

Inventor of the phonograph

I NEVER THINK OF THE PAST. I GO TO SLEEP THINKING ONLY OF WHAT I AM GOING TO DO TOMORROW.

George Westinghouse (1846–1914)

Inventor of the railway air brake

LET US GO ON EXECUTING THE THINGS WE UNDERSTAND AND ***LEAVE THE REST TO YOUNGER MEN,*** *WHO HAVE NEITHER MONEY NOR CHARACTER TO LOSE.*

James Watt (1736–1819)

Inventor of the Watt steam engine

Everybody who ages is going to be their own problem-solver.

Barbara Beskind (1924–)

Inventor of inflatable equipment for teaching children balance

I believe that scientists get their best work done before they are fifty, or even earlier than that. **I did most of my best work while I was young.**

Claude Shannon (1916–2001)

Pioneer of information theory

Don't think you're too old to do things. You're never too old.

Pansy Ellen Essman (1918–2011)

Inventor of the Pansy-ette infant bath cushion

We live in a world changing so rapidly that what we mean frequently by common sense is ***doing the thing that would have been right last year.***

Edwin Land (1909–91)

Inventor of the Polaroid camera

YOU KNOW WHAT THE DIFFERENCE IS BETWEEN SCIENCE FICTION AND SCIENCE? TIMING.

Dean Kamen (1951–)

Inventor of the Segway

You have to be more than yourself.
The world is more important than you are.
Having a view of the world that is larger than you helps to maintain a productive, positive, expanded view.

Barbara Beskind (1924–)

Inventor of products and technologies for the elderly

ANY SOIL IS A COUNTRY TO THE BRAVE, AND THE HEAVENS ARE EVERYWHERE OVERHEAD.

Tycho Brahe (1546–1601)

Inventor of the Tychonian quadrant

Anybody can be an inventor. Sometimes you ask yourself "Why doesn't someone come up with a better idea?" Well, why don't you do it?

Emanuel Logan Jr. (1933–2005)

Inventor of the Door Guard emergency exit lock

Adler, Robert **114**
Aliyu, Jelani **48**
Ando, Momofuku **57**
Babbage, Charles **11, 112**
Baylis, Trevor Graham **101**
Bell, Alexander Graham **22, 38, 127**
Berners-Lee, Sir Tim **44, 98**
Beskind, Barbara **147, 152**
Bessemer, Sir Henry **16**
Birdseye, Clarence **26**
Bohr, Niels **128**
Borden, Gail **83, 95**
Brahe, Tycho **153**
Brandenburg, Karlheinz **45**
Brattain, Walter **113**
Bushnell, Nolan **7, 61, 125**
Camras, Marvin **105**
Carlson, Chester **14, 21, 100**
Carver, George Washington **33**
Christie, Agatha **62**
Coston, Martha J. **94**
Crutchfield, E. Bryant **60**
da Vinci, Leonardo **47, 54, 84**
Damadian, Raymond **116**
Darden, Christine **93**
Eckert, J. Presper **110**
Edison, Thomas **65, 103, 133, 144**
Engelbart, Douglas **104, 117**
Essman, Pansy Ellen **149**
Faraday, Michael **8, 82**
Ford, Henry **74, 121**
Franklin, Benjamin **39**
Fry, Arthur **31**
Fuller, R. Buckminster **80**
Fulton, Robert **72, 99, 129**
Gabor, Dennis **34**
Galilei, Galileo **40, 120**
Gillette, King Camp **25**
Gould, Gordon **96**
Graham, Bette Nesmith **58**
Grandin, Temple **69**
Greatbatch, Wilson **71**
Gundlach, Robert **97**
Hall, Charles Prior **86**
Handler, Ruth **32**
Henry, Beulah Louise **70**
Hibbard, Susan **59**
Hodgkin, Dorothy Crowfoot **124**
Hoff, Marcian **134**
Holmes, Arthur **119**
Hooke, Robert **46**
Hopper, Grace Murray **42, 91, 136**
Jarvik, Robert **18**

Jobs, Steve **160**
Johnson, Lonnie **19**
Kamen, Dean **151**
Kelly, Mervin **77**
Kettering, Charles **142**
Kilby, Jack **28**
Knight, Margaret **13**
Kurzweil, Raymond **76**
Kwolek, Stephanie **12**
Land, Edwin **41, 132, 150**
Logan, Emanuel, Jr. **90, 154**
Lovelace, Ada **126**
MacCready, Paul **20, 66**
Mangano, Joy **37**
Moreno, Roland **85, 131**
Morita, Akio **36**
Musk, Elon **139**
Naismith, James **23**
Newton, Sir Isaac **92, 118**
Oppenheimer, J. Robert **111**
Ovshinsky, Stanford **75**
Perkin, William Henry **30**
Pincus, Gregory Goodwin **135**
Polley, Eugene **89**
Rabinow, Jacob **78**
Rohwedder, Otto Frederick **88**
Rubik, Ernő **9, 56**
Seacat DeLuca, Lisa **137**
Sejong the Great **49**
Senefelder, Alois **15, 107**
Shannon, Claude **148**
Sims, Tom **52**
Spaeth, Mary **17**
Sundback, Gideon **53**
Szpur, Roman **81**
Tesla, Nikola **64, 68, 79**
Thomson, Sir William (Lord Kelvin) **122**
Tupper, Earl Silas **24**
Turing, Alan **138**
Twain, Mark **63**
von Neumann, John **123**
von Zeppelin, Ferdinand **50**
Walker, Edward Craven **87**
Walker, Madame C. J. **102, 130**
Wang, An **27, 43, 143**
Watson-Watt, Sir Robert **51, 67**
Watt, James **109, 146**
Westinghouse, George **35, 145**
Whitney, Eli **106, 108**
Wozniak, Steve **29, 73**
Wright, Orville **10**
Wright, Wilbur **55**
Wyeth, Nathaniel **141**
Zuckerman, Ethan **115**
Zworykin, Vladimir K. **140**

ACKNOWLEDGMENTS

This book simply would not exist without the diligent efforts of Simone Kaplan-Senchak and Sara Stemen. In addition to researching, contributing, and verifying quotes, they invented a system for organizing them and, in Sara's case, made the first pass in grouping them into the clever pairings you find here. To them, as well as all my colleagues at the Press who sent their favorite quotes and make publishing and writing such a rewarding experience, my heartfelt thanks.

Kevin Lippert
Ghent, New York
December 2016

Published by
Princeton Architectural Press
A McEvoy Group company

37 East 7th Street
New York, New York 10003

202 Warren Street
Hudson, New York 12534

www.papress.com

Printed and bound in China
20 19 18 17 4 3 2 1
First edition

Editor: Sara Stemen
Research assistance: Simone
Kaplan-Senchak, Nolan Boomer,
Esme Savage, Faith Shaeffer
Designer: Benjamin English
Series designer: Paul Wagner

Special thanks to:
Janet Behning, Nicola Brower,
Abby Bussel, Tom Cho,
Barbara Darko, Jenny Florence,
Jan Cigliano Hartman,
Susan Hershberg, Lia Hunt,
Mia Johnson, Valerie Kamen,
Jennifer Lippert, Kristy Maier,
Sara McKay, Eliana Miller,
Wes Seeley, Rob Shaeffer,
and Joseph Weston of Princeton
Architectural Press
—Kevin C. Lippert, publisher

Library of Congress
Cataloging-in-Publication Data

Names: Lippert, Kevin C., editor.
Title: The inventor says :
quotes, quips, and words of
wisdom / compiled & edited by
Kevin Lippert.
Description: New York : Princeton
Architectural Press, [2017] |
Series: Words of wisdom | Index
of inventors.
Identifiers: LCCN 2016059212 |
ISBN 9781616896225
Subjects: LCSH: Inventors —
Quotations, maxims, etc.
Classification: LCC T47 .I58
2017 | DDC 080—dc23
LC record available at
https://lccn.loc.gov/2016059212

The only way to do great work is to love what you do. If you haven't found it yet, keep looking.

Steve Jobs (1955–2011)

Cofounder and CEO of Apple Inc.